Reasoning and Applied MATHEMATICS for the Early Years

Reasoning and Applied MATHEMATICS for the Early Years

A Handbook for Teachers

Robert L. Hammond,
B.A., M.A., Ed.D.

iUniverse, Inc.
New York Lincoln Shanghai

REASONING AND APPLIED MATHEMATICS FOR THE EARLY YEARS

A Handbook for Teachers

iUniverse books may be ordered through booksellers or by contacting:

iUniverse
2021 Pine Lake Road, Suite 100
Lincoln, NE 68512
www.iuniverse.com
1-800-Authors (1-800-288-4677)

91725 WINNEBAGO ST.
EUGENE, OREGON 974O8
Dissertation 1962
Program Evaluation-Handbook 1—Independent Basic Skill Development
LIBRARY OF CONGRESS CERTIFICATE OF COPY RIGHT REGISTRATION
REGISTRATION NUMBER TXu 247 226 1986

ISBN-13: 978-0-595-40784-2 (pbk)
ISBN-13: 978-0-595-85148-5 (ebk)
ISBN-10: 0-595-40784-6 (pbk)
ISBN-10: 0-595-85148-7 (ebk)

Printed in the United States of America

CONTENTS

PART I

INTRODUCTION AND EXPLANATION

Foreword

Beginning in elementary and middle school, students are taught arithmetic based on rote learning. Rote learning continues with the introduction to pure mathematics. The National Council of Teachers of Mathematics is calling for a quick response to basic operations in the elementary grades. A quick response can best be achieved with reasoning as the major focus of learning.

Rote leaning without reasoning contradicts Einstein's concept of number. In his *Remarks on Bertrand Russell's Theory of Knowledge,* he says, "The concept of number is a free creation of thought, a self-created tool which simplifies the organizing of sensory experience, but a tool which cannot inductively be gained from sense experience." To apply arithmetic/mathematics in the physical world is to bring creative logic to bear on observed facts.

As a beginning teacher in a small school district in California, I had the opportunity to teach split grades four and five. I worked with parents and served on committees for parent instruction. Parents had many unanswered questions as to why their children were having trouble with arithmetic and reading. I had the same concerns. The literature, at that time, gave no answers that would help students and parents.

I moved to Arizona where I taught grades six and seven. I worked with Caucasian, Indian, Black and Hispanic children. Problems in reading, arithmetic, science and social studies still plagued me. Students could read well orally but had problems applying what they had read.

The following year I was invited to teach in the lab school at Northern Arizona State College. During my second year, I was assigned a course to help future teachers improve their math skills. After working with the teachers for a semester, it became clear that the relationship between reasoning and applied mathematics was a problem. The class traced the problem back to rote learning in the elementary grades.

Research

I entered the University of Southern California as a graduate, teaching in the College of Education. I pursued a study on the nature of objects and mathematics. My conclusion was that any attempt to examine reasoning must be approached through the four basic operations of arithmetic and mathematics. This will be discussed in chapter one.

University and School Experience

My teaching at the University of British Columbia, the University of Arizona and Ohio State and my development of, and participation in evaluation centers, was useful but sidetracked me from my search for reasoning. While teaching at the University of Arizona I received grants for the development of one of three evaluation centers (EPIC) in the United States. After leaving Arizona I accepted a position at The Ohio State University, where I was the co-director of the evaluation center and director of the test development center. I also served on the Phi Delta Kappa Study Committee which wrote the book "*Educational Evaluation and Decision Making.*" After leaving Ohio State I worked for the Montana State Department of Education, The University of Oregon and the Springfield School District from which I retired in 1994. My career was devoted to evaluation and testing. I now have the time to pursue reasoning and applied mathematics.

CHAPTER I

Four Basic Operations of Mathematics

Fundamental to all areas of mathematics are the basic operations of addition, multiplication, subtraction and division. The principles that govern these operations help to describe the elements of a system of mathematics by prescribing their behavior under certain conditions. The principles governing the operations tell all that the elements can do; all their properties are consequent from these fundamental rules of behavior.

A mathematical operation is governed by certain signs called the objects of mathematics. In any discussion of the objects of mathematics we must first carefully separate, in our minds, mathematics and the mathematical treatment of certain entities. Thus, it appears that there are two distinct approaches to operations. For the purpose of applied mathematics the approach to understanding of the operations is thought of in terms of mathematical application, not pure mathematics.

The author's dissertation "**Ability With The Mathematical Principles Governing the Operations of Addition, Multiplication, Subtraction and Division**" received the Alumni Award for Outstanding Doctoral Research in the School of Education at the University of Southern California in 1962.

The scope of this study was to ascertain the understanding by 300 seventh grade children of certain mathematical principles governing the operations of addition, multiplication, subtraction and division, and the relationship of this understanding to arithmetic and mental ability as measured by standardized tests.

A 50 item test was developed for this study. A description of each operation in terms of the axiomatic (a proposition regarded as a self-evident truth) method was developed for the study. There were five items for each proposition.

Propositions

The principles governing the propositions for closure, sequential order, grouping, axiom of subtraction, definition of difference, axiom of division and definition of quotient were applied to the mathematical situations. Set I responses were taken directly from the 50 item test used in the study. The principles are illustrated below.

CRITICAL READING

For each set (I, II, III) there is one or more correct answers to the story below. Which of the combinations gives us the same information found in the story? Place an x on the line after the response you select. If none of the combinations gives us the same information, then select the number 5 labeled none.

Closure property

5. Paul started with e airplanes. He made d more airplanes. He had e + d airplanes in all.

CRITICAL THINKING

	I		II		III	
1.	e + d = e x d	_____	d/e = e + d	_____	e +d = e-d	_____
2.	d = e + d	_____	x - d = e	_____	e/d = e + d	_____
3.	e - d = e + d	_____	e + d = e - d	_____	x - e = d	_____
4.	e + d = e + d.	__x__	e + d = e/d	_____	e + d = d/e	_____
5.	none	_____	none	_____	none	_____

Sequential Order

11. George had a white mice. He sold all of them at b cents each. George received a x b cents for the mice.

CRITICAL THINKING

	I		II		III	
1.	a / b = b - a	____	b x a = b x a	____	$\frac{a \times b}{}$ = a	____
2.	a x b = a - b	____	a - b = a x b	____	a + b = a - b	____
3.	a + b = a x b	____	a/b = a/b	____	$\frac{a \times b}{a}$ = b	____
4.	a x b = b x a	x	a - b = a - b	____	a x b = a x b	____
5.	none	____	none	____	none	____

Grouping

14. Mother bought a melon for b cents and cookies for e cents and eggs for f cents. Mother spent b + e + f cents in all.

CRITICAL THINKING

	I		II		III	
1.	(b + e) + f = b + (e + f)	x	b + e + f = f + e + b	___	b x f = e	___
2.	b x e x f = b + e + f	___	b + e + f = b - e - f	___	b - e = f	___
3.	(b + e) - f = b + e + f	___	b +e + f = f x b	___	(b +e) - f = b + e + f	___
4.	b - e + f = b + e + f	___	b + f - e = b + f	___	b + e - f = b + f + e	___
5.	none	___	none	___	none	___

The basic principles involved were used in the same manner for both addition and multiplication.

Axiom of subtraction

29. Mother baked f cherry tarts. The family ate e tarts for lunch. There were z tarts left.

CRITICAL THINKING

	I		II		III	
1.	z - f = e	____	d + z + e = f + z + e	____	z / f = e	____
2.	f x z = e	____	e + z = f	____	f - z = e	____
3.	z + e = f	x	z x f = z x f	____	z x e = f	____
4.	e x f = z	____	e - f = z	____	f / e = z	____
5.	none	____	none	____	none	____

Definition of difference

4. Dick's allowance is a cents a week. Jim's is b cents. Jim's allowance is z cents less than Dick's.

CRITICAL THINKING

	I		II		III	
1.	b x z = a	____	z - a = b	____	a/ b = z	____
2.	a + b = z	____	b x a =	____	z + a = b	____
3.	a - b = z	x	z + b = a	____	z /a = b	____
4.	b - z = a	____	a x z = b	____	c / a = z	____
5.	none	____	none	____	none	____

Axiom of division

41. There are b desks in Bill's room. There are c desks in each row. There are z rows in the room.

CRITICAL THINKING

	I		II		III	
1.	b + c = z	____	b /c = z	____	z = c - b	____
2.	z - c = b	____	z x c = b	____	z + b = c	____
3.	c x z = b	x	c / b = z	____	b /z = c	____
4.	c x b = z	____	z / b = a	____	z / b = c	____
5.	none	____	none	____	none	____

Definition of quotient

23. a boys shared equally the c dollars they earned. Each boy received z dollars.

CRITICAL THINKING

	I		II		III	
1.	a + c = z	_____	a x z = c	_____	a x c = z	_____
2.	a - c = z	_____	a / c = z	_____	z - c = a	_____
3.	c/a = z	X	z - a = c	_____	a + z = z	_____
4.	z/a = c	_____	a x a = z	_____	z x a = c	_____
5.	none	_____	none	_____	none	_____

Applications for Educational Change Produced by This Study

1. The seventh-grade child is capable of mathematical reasoning on the more advanced level described by the axiomatic method of mathematics. On this basis, the instructional program should include material designed to develop this ability.
2. Instruction programs for mathematical reasoning should be provided for all levels of ability, with ample opportunities for the more capable students to advance at a rate determined by their individual abilities.
3. The ability to apply the principles governing the operations of subtraction and division appear to show a close relationship between factors of IQ, arithmetic achievement and algebra aptitude. To care for the wide ranges of differences in ability, there should be a differentiated program of instruction in the principles governing these operations.
4. Reasoning in terms of applied mathematics is one method that appears to be successful for providing the techniques necessary for describing the mathematical principles governing the operations.

5. The applied mathematical situation does not need to be highly complicated to determine mathematical reasoning ability.
6. Mathematical principles governing the operations can derive their meaning from concrete experiences. Therefore, it would seem that any instructional program should capitalize on concrete experiences to develop the principles involved in the operations.

CHAPTER II

Reasoning

> Reasoning—"The drawing of inferences or conclusions through the use of reason..."
> Reason—..."2a The power of comprehending, or thinking esp. in orderly rational ways..."[1]

A student participates in instructional programs for language arts, mathematics and science. The ultimate goal in each instructional area is to exercise the power of reasoning to gain new knowledge. Reasoning, in this context, is a developmental process involving skills that are not unique to any one area but do involve content setting that place greater demands on the refinement of such skills as they progress from one grade level to the next.

Reasoning, as a process, involves three performance levels that are interdependent and sequential. The process begins with critical reading, critical thinking and ends with critical writing. The first level, reading, starts the process with recognition, understanding and usage. The second level, thinking, begins within an examination of patterns and structures. The third level ends the process with writing involving quantitative or qualitative judgments.

---

1. Webster's New Collegiate Dictionary, Springfield. Mass. : G & C Merriam Co. 1975, p. 62

Teacher participation

Read the following directions: As a teacher you are going to work a new and interesting kind of arithmetic puzzle. The puzzles are easy because the answers are given.

CRITICAL READING

For each set (I, II, III) there is one or more correct answers. Which of the combinations gives us the same information found in the story? Place an x on the line after the response you select. If none of the combinations gives us the same information, then select the number 5 labeled "none."

In the space provided tell why the items you marked are correct for set I. Use the words and symbols provided.

24. The flying distance from Chicago to Denver is e miles. The Jones family made the trip in f hours. The plane averaged z miles per hour.

CRITICAL THINKING

	I		II		III	
1.	f x e = z	_____	e - f = z	_____	f x e = z	_____
2.	f / e = z	_____	f - e = z	_____	e x f = z	_____
3.	e / f = z	_____	e + f =	_____	f x z = e	_____
4.	e + f = z	_____	z x f = e	_____	z. x e = f	_____
5.	none	_____	none	_____	none	_____

____

____

____

____

____

7. Fred saved b dollars in c weeks. He saved w dollars each week.

	I		II		III	
1.	w x b = c	_____	w - c = b	_____	b / w = c	_____
2.	b + c = w	_____	w / b = c	_____	c x w = b	_____
3.	b / c = w	_____	w x c = b	_____	c - w = b	_____
4.	c / w = b	_____	b + c = w	_____	c / b = w	_____
5.	none	_____	none	_____	none	_____

____

____

____

____

____

Student Participation

The above exercises were administered to 124 eighth grade students in April of 2006. Twenty-five percent of the students were able to understand some part of the exercise. Two students had all the answers correct. The students were required to develop a written response for set I. It was obvious the students were frustrated with the writing section.

Let us expand our dictionary definition for reasoning. There are two types of reasoning defined by the University of Toronto Mathematics Network: 1. "Inductive reasoning is part of the discovery process. The observations of special cases leads one to suspect very strongly (though not with absolute logical certainty) that some general principle is true." 2. "Deductive reasoning is the method you use to demonstrate with logical certainty the principle is true."

Both are a necessary part of applied mathematical reasoning. Reviewing the math situation you find an item you think is correct. Then, once your suspicions have given you a target and a direction for your deductive reasoning, you construct your rigorous logical response using deductive reasoning. This all starts with critical reading, then critical thinking and ends with critical writing. We will use student responses to demonstrate the process.

24. The flying distance from Chicago to Denver is e miles.

The Jones family made the trip in f hours. The plane averaged z miles per hour.

Student 1

	I		II		III	
1.	f x e = z		e - f = z		f x e = z	
2.	f / e = z		f - e = z		e x f = z	
3.	e / f = z	x	e + f =		f x z = e	x
4.	e + f = z		z x f = e	x	z x e = f	
5.	none		none		none	

"In set one I thought of different kinds of equations to get the answer which was e/f = z. This made sense because the distance to Chicago divided by how many hours it took equals the average speed of the plane."

Student 2

	I		II		III	
1.	f x e = z		e - f = z		f x e = z	
2.	f / e .= z		f - e = z		e x f = z	
3.	e / f = z	x	e + f =		f x z = e	? .
4.	e + f = z		z x f = e	?	z x e = f	
5.	none		none	x	none	x

"The equation e/f = z is correct for set one because the distance from Chicago to Denver is e, the trip took f hrs. and the plane's average is z miles (e/f = z).To find the average speed you divide total distance by

total time. If the numbers were listed you would be able to find the average miles per hr."

27. Fred saved b dollars in c weeks. He saved w dollars each week.

Student 1

	I		II		III	
1.	w x b = c		w - c = b		b / w = c	x
2.	b + c = w		w / b = c		c x w = b	?
3.	b / c = w	x	w x c = b	x	c - w = b	
4.	c / w = b		b + c = w		c / b = w	
5.	none		none		none	

"In set one I picked #3 which is b / c = w. This was right. The amount just saved divided by how many weeks equal how much money he saved in a week."

Student 2

	I		II		III	
1.	w x b = c		w - c = b		b / w = c	?
2.	b + c = w		w / b = c		c x w = b	?
3.	b / c = w	?	w x c = b	?	c - w = b	
4.	c / w = b		b + c = w		c / b = w	x
5.	none	x	none	x	none	

"The correct answer for 27 is none because none of the equations will work for the problem. c/b = w weeks divided by dollars will equal the dollars each week."

Now we will take a look at reasoning as a process. Both inductive and deductive reasoning involves the same process components. The process components are critical reading, critical thinking and critical writing.

Inductive reading—Both students had to read the proposition statements and review the multiple choice items.

Inductive thinking—After the first review, they looked at alternatives through a second reading.

Inductive writing—Any writing that occurred could be called experimenting with combinations.

Deductive reading—A final reading to confirm conclusion

Deductive thinking—A review to select multiple choice items

Deductive writing—A written explanation for why the item is correct, giving us insight into the process of selection.

Problem 24—both students had set I correct. Everything broke down for student 2 in sets II and III. The student was locked into division and could not distinguish the relationship with multiplication. The student was very articulate. We are unable to say if it was a case of not knowing or recognizing the relationship or faulty reasoning.

Problem 27—student 1 had the correct answer for set I. In set III the student appeared to not recognize the concept that c x w = b is the same as w x c = b. Based on this student's performance, I would suggest this may be carelessness or trying to hurry to finish before others. Student 2 did not get any of the answers correct. We would have to discuss this problem with the student to determine the cause.

A small mixed group of students, grades 9 through 12, did the exercise the summer of 2006. One eleventh grade student got them all correct, but used algebra and number expressions. The student did not give a verbal explanation.

One twelfth grade student managed to get one item correct and made the following comment:

> *"Since I'm not very good at math, it's taken a sheer act of God that I've gotten this far. I don't really understand what is going on. I just hope I got everything right!"*

One ninth grade student managed to get four of the items correct and made the following comment:

> *"I have no clue. I just guess that was the correct answer by my own knowledge of what I have already learned."*

The majority of students, teachers and the general population have been so conditioned to symbol and number memorization that they do not use the power of reasoning to find solutions involving proposition-like statements.

CHAPTER III

Blueprint for Instruction

Any attempt to program instructors, or indeed oneself, to implement instruction is likely to arouse an indignant resistance. This is true of most teachers. The resistance to such force is even greater when formulas are produced from the behavioral sciences, which rely heavily on common sense and reasoning.

This handbook does not attempt to dictate instructional methods for creating a learning environment for students. It should be looked at as one set of many tools to be used by the teacher. The best source of information is to study students' responses to the critical writing phase of the reasoning process. More can be learned from your students than any other source as to how instruction should be organized. A plan will be suggested, so use the parts that best fit your approach to teaching.

Instructional Blueprint

This instructional blueprint is based on a common sense approach to instruction through the behavioral sciences and facet design. A synthesized statement from "*Field Theory in Social Science*" by Kurt Lewin states, "In principle, it is everywhere accepted that behavior (B) is a function (F) of people (P) and their environment (E). Therefore, B = F (P, E) = (LSp). The life space (LSp) includes the person and their psychological learning environment.

There are two steps to facet design:

1. Definition of the basic set of terms or elements that lead to facets; and
2. Definition of a new set of elements which is the cartesian product of the facets. Each element of the set is a combination of the facets and is called the variables of the design.

The behavior facet (B) is defined by two elements:

B = B1 inductive reasoning
B2 deductive reasoning

B1 inductive reasoning—Inductive reasoning is part of the discovery process. The observations of special cases lead one to suspect very strongly (though not know with absolute certainty) that some general principle is true.

B2 deductive reasoning—Deductive reasoning is the method used to demonstrate with logical certainty that the principle is true.

The people facet (P) is defined by one element:

P = P1 students

P1 Students—refers to students in grades four through eight.

The instructional environment facet (E) has three elements

E = E1 critical reading
E2 critical thinking
E3 critical writing

E1 Critical reading—Once we have determined that a proposition is consistent and coherent we can begin to develop conclusions.

E2 Critical thinking—We must decide what to accept as true and useful.

E3 Critical writing—We must support our conclusion with detailed evidence from the proposition under examination. Do not forget to document elements and phrases within the proposition.

Blueprint for instruction

(B1 inductive reasoning)
The instructional behavior (B) { } for
(B2 deductive reasoning)
Students (P1) in the instructional skills environment (E)
(E 1) critical reading
{(E2) critical thinking)} yield a range (R) of performances for
(E3 critical writing)
applied mathematical situations while teaching reading, thinking, writing and the process of reasoning for students with above average, average and below average skill development

What Do I Do for Students Who are Low Achievers?

The first task is to build the students' confidence in the learning process. Students need to be successful. To achieve success they must work with a problem they have no trouble understanding. The teacher must work with them at this stage until they are successful. Give the students time to develop their responses. Read the problem with the students. Discuss what is meant by critical reading, critical thinking and critical writing.

CRITICAL READING

Which of the combinations below gives us the same information found in the story? If none of the combinations gives us the same information, select number 5 labeled "none."

2. Jane earned a cents. She put b cents of this money in the bank. She had z cents left.

CRITICAL THINKING

1. a/b = z
2. a +b = a
3. b - z = a
4. a - b = z
5. none

CRITICAL WRITING

____

____

____

____

Let us look at each of the possible answers to the puzzle. Look at each answer very carefully to avoid careless mistakes.

Number 1 tells us that if we divide a (the money Jane earned) by b (the money she put in the bank) we will get z (the money she had left).

Number 2 tells us that if we add a (the money Jane earned) and b (the money she put in the bank) we will get z (the money Jane had left).

Number 3 tells us that if we subtract z (the money Jane had left) from b (the money she put in the bank) we will get a (the money she earned).

Number 4 tells us that if we subtract b (the money Jane put in the bank) from a (the money she earned) we get z (the money she had left).

Number 5 tells us that none of the above answers is correct.

Do not move on to puzzle number two until the students have mastered puzzle number one. Take plenty of time, the students must gain confidence before they move to the student handbook.

CRITICAL READING

Which of the combinations gives us the same information found in the story? If none of the combinations give us the same information, then select number 5 labeled "none."

3. Dick's mother gave him d cents. He spent e cents for gum. Dick had w cents left

CRITICAL THINKING

1. $d/e = w$
2. $e - d = w$
3. $d + e = w$
4. $w + e = d$
5. none

CRITICAL WRITING

____

____

____

____

____

____

____

Conclusions

As a teacher, you have been given an approach to improve student reasoning, applied mathematics, critical reading, critical thinking and critical writing based on research and practical implementation. The student workbook that follows gives you examples covering ten mathematical principles, each requiring ample opportunity for instruction and practice. A design for instruction has been suggested that gives you all the flexibility for methods instruction that you will ever need to fit your style of teaching. The student handbook may be ordered separately for each student.

STUDENT REASONING WORKBOOK

CRITICAL READING

Read each story and set of answers very carefully. The puzzles are easy, but it is still possible to make careless mistakes.

CRITICAL THINKING

Place an X in the space provided for each correct answer in each of the multiple choice sets (I, II, III) that fit the verbal statement.

CRITICAL WRITING

Develop a written statement telling why you marked the item correct (using each letter assigned as a value). Include the response "none" as an item in the written statements.

CRITICAL READING

For each set (I, II, III) there is one or more correct answers to the story below. Which of the combinations gives us the same information found in the story? Place an x on the line after the response you select. If none of the combinations gives us the same information, then select the number 5 labeled none.

In the space provided tell why the items you marked in set I, II and III are correct. Use the words and symbols in the story.

1. Tom started with a bag of d marbles. He bought e more marbles. Tom had x marbles in all.

CRITICAL THINKING

	I		II		III	
1.	d - e = d + e	_____	d + e = d - e	_____	e + d = d - e	_____
2.	d + e =d + e	_____	x + e = d	_____	x - e = d	_____
3.	d + e = d	_____	e + d = x	_____	d - e = d + e	_____
4.	d/e = d +e	_____	d x e = d + e	_____	d - x = e	_____
5.	none	_____	none	_____	none	_____

CRITICAL WRITING

____

____

____

____

____

CRITICAL READING

For each set (I, II, III) there is one or more correct answers to the story below. Which of the combinations gives us the same information found in the story? Place an x on the line after the response you select. If none of the combinations gives us the same information, then select the number 5 labeled none.

In the space provided tell why the items you marked in set I, II and III are correct. Use the words and symbols in the story.

2. Larry saved d dollars to spend for vacation fun. First he bought a basketball for e dollars. Larry had w dollars left.

CRITICAL THINKING

	I		II		III	
1.	e - w = d	_____	w + e = d	_____	w/e =c	_____
2.	w x e = d	_____	e x w = d	_____	w/c w = d	_____
3	e x d = w	_____	d x e = w	_____	e + w = d	_____
4.	d - e = w	_____	w - d = e	_____	e x w x d = w - d	_____
5.	none	_____	none	_____	none	_____

CRITICAL WRITING

____

____

____

____

____

____

____

____

CRITICAL READING

For each set (I, II, III) there is one or more correct answers to the story below. Which of the combinations gives us the same information found in the story? Place an x on the line after the response you select. If none of the combinations gives us the same information, then select the number 5 labeled none.

In the space provided tell why the items you marked in set I, II and III are correct. Use the words and symbols in the story.

3. There are a boys and b girls in Jack's room. There are x children in Jack's room.

CRITICAL THINKING

	I		II		III	
1.	a - b = a x b	_____	a + b = a + a	_____	a / b = c	_____
2.	a x b = a + b	_____	a x b = a- x b	_____	a/b = d	_____
3.	a + b = c	_____	a + b = x	_____	b x a = x	_____
4.	a + b = a - b	_____	c /c = a	_____	a + b = a + b	_____
5.	none	_____	none	_____	none	_____

CRITICAL WRITING

____

____

____

____

____

____

____

____

CRITICAL READING

For each set (I, II, III) there is one or more correct answers to the story below. Which of the combinations gives us the same information found in the story? Place an x on the line after the response you select. If none of the combinations gives us the same information, then select the number 5 labeled none.

In the space provided tell why the items you marked in set I, II and III are correct. Use the words and symbols in the story.

4. Dick's allowance is a cents a week. Jim's is b cents. Jim's allowance is z cents less than Dick's.

CRITICAL THINKING

	I		II		III	
1.	b x z = a	_____	z - a = b	_____	a/ b = z	_____
2.	a + b = z	_____	b x a =	_____	z + a = b	_____
3.	a - b = z	_____	z + b = a	_____	z /a = b	_____
4.	b - z = a	_____	a x z = b	_____	c / a = z	_____
5.	none	_____	none	_____	none	_____

CRITICAL WRITING

____

____

____

____

____

____

____

____

CRITICAL READING

For each set (I, II, III) there is one or more correct answers to the story below. Which of the combinations gives us the same information found in the story? Place an x on the line after the response you select. If none of the combinations gives us the same information, then select the number 5 labeled none.

In the space provided tell why the items you marked in set I, II and III are correct. Use the words and symbols in the story.

5. Paul started with e airplanes. He made d more airplanes. He had e + d airplanes in all.

CRITICAL THINKING

	I		II		III	
1.	e + d = e x d	_____	d/e = e + d	_____	e +d = e-d	_____
2	d = e + d	_____	x - d = e	_____	e/d = e + d	_____
3.	e - d = e + d	_____	e + d = e - d	_____	x - e = d	_____
4.	e + d = e + d	_____	e + d = e/d	_____	e + d = d/e	_____
5.	none	_____	none	_____	none	_____

CRITICAL WRITING

____

____

____

____

____

____

____

____

CRITICAL READING

For each set (I, II, III) there is one or more correct answers to the story below. Which of the combinations gives us the same information found in the story? Place an x on the line after the response you select. If none of the combinations gives us the same information, then select the number 5 labeled none.

In the space provided tell why the items you marked in set I, II and III are correct. Use the words and symbols in the story.

6. At the Lake School, there are b children. There are a boys and z girls.

CRITICAL THINKING

	I		II		III	
1.	a + z = b	_____	b/a = z	_____	z - a = b	_____
2.	b x a = z	_____	b x a = c	_____	b - a = a	_____
3.	z x b = a	_____	b - z = a	_____	z/a = b	_____
4.	z + b = a	_____	c/a = b	_____	a - b = z + b	_____
5.	none	_____	none	_____	none	_____

CRITICAL WRITING

____

____

____

____

____

____

____

____

CRITICAL READING

For each set (I, II, III) there is one or more correct answers to the story below. Which of the combinations gives us the same information found in the story? Place an x on the line after the response you select. If none of the combinations gives us the same information, then select the number 5 labeled none.

In the space provided tell why the items you marked in set I, II and III are correct. Use the words and symbols in the story.

7. In a dart game, Nick made f points, g points and h points. Nick made f + g + h in all.

CRITICAL THINKING

	I		II		III	
1.	f + g + h = f x g x h	___	g + h + f = h + f + g	___	f/g = h	___
2.	f + g = h	___	h/g = f	___	f x g = f - g - h	___
3.	f + g + h = f +g + h	___	f x g x h = f x g x h	___	f - g - h = f - g	___
4.	f - h = f + g + h	___	g / h = f	___	h + g = h + g	___
5.	none	___	none	___	none	___

CRITICAL WRITING

___

___

___

___

___

___

___

___

CRITICAL READING

For each set (I, II, III) there is one or more correct answers to the story below. Which of the combinations gives us the same information found in the story? Place an x on the line after the response you select. If none of the combinations gives us the same information, then select the number 5 labeled none.

In the space provided tell why the items you marked in set I, II and III are correct. Use the words and symbols in the story.

8. Jim had a marble shooters and b other marbles. Jim had a + b marbles of both kinds.

CRITICAL THINKING

	I		II		III	
1.	a = a + b	_____	b + a = b + a	_____	a - b = b/a	_____
2.	a - b = a + b	_____	a + b + a = b	_____	a / b = a + b	_____
3.	a + b = a + b	_____	a + b = b - a	_____	b + a = a + b	_____
4.	a + b = a x b	_____	ax b = a / b	_____	a - b = a - b	_____
5.	none	_____	none	_____	none	_____

CRITICAL WRITING

____

____

____

____

____

____

____

____

CRITICAL READING

For each set (I, II, III) there is one or more correct answers to the story below. Which of the combinations gives us the same information found in the story? Place an x on the line after the response you select. If none of the combinations gives us the same information, then select the number 5 labeled none.

In the space provided tell why the items you marked in set I, II and III are correct. Use the words and symbols in the story.

9. At summer camp c boys can sleep in each tent. There are f tents. c x f boys can sleep in tents at the camp.

CRITICAL THINKING

	I		II		III	
1.	f/c = c x f	_____	c/f = c x fv	_____	c + f = c - f	_____
2.	c - f = c x f	_____	c x f = f x c	_____	c x f = f /c	_____
3.	c x f = c x f	_____	c/f = f/c	_____	c/f x f = c x f	_____
4.	c + f = c x f	_____	f/c + c/f = f x c	_____	f x c = c x f	_____
5.	none	_____	none	_____	none	_____

CRITICAL WRITING

____

____

____

____

____

____

____

____

CRITICAL READING

For each set (I, II, III) there is one or more correct answers to the story below. Which of the combinations gives us the same information found in the story? Place an x on the line after the response you select. If none of the combinations gives us the same information, then select the number 5 labeled none.

In the space provided tell why the items you marked in set I, II and III are correct. Use the words and symbols in the story.

10. At the school cafeteria, Billy bought a sandwich for c cents; he also bought a bowl of soup for d cents and milk for e cents. Billy spent c + d +e cents at the cafeteria.

CRITICAL THINKING

	I		II		III	
1.	c x d x e = c +d +e	____	c x d =e	____	e x d = c	____
2.	d + c + e = e + c +d	____	c / e = d	____	e + d +c = d + e + c	____
3.	c - d - e = c + d + e	____	$\frac{c \times d}{e} = d \times e$	____	d - c= e	____
4.	c + d +e= d	____	$\frac{c + e - d}{e} = d \times e$	____	c + d x e = c x d	____
5.	none	____	none	____	none	____

CRITICAL WRITING

____

____

____

____

____

____

CRITICAL READING

For each set (I, II, III) there is one or more correct answers to the story below. Which of the combinations gives us the same information found in the story? Place an x on the line after the response you select. If none of the combinations gives us the same information, then select the number 5 labeled none.

In the space provided tell why the items you marked in set I, II and III are correct. Use the words and symbols in the story.

11. George had a white mice. He sold all of them at b cents each. George received a x b cents for the mice.

CRITICAL THINKING

	I		II		III	
1.	a x b = b - a	_____	b x a = b x a	_____	$\frac{a \times b}{}$ = a	_____
2.	a / b = a - b	_____	a - b = a x b	_____	a + b = a - b	_____
3.	a + b = a x b	_____	a/b = a/b	_____	$\frac{a \times b}{a}$ = b	_____
4.	a x b = b x a	_____	a - b = a - b	_____	a x b = a x b	_____
5.	none	_____	none	_____	none	_____

CRITICAL WRITING

____

____

____

____

____

____

____

CRITICAL READING

For each set (I, II, III) there is one or more correct answers to the story below. Which of the combinations gives us the same information found in the story? Place an x on the line after the response you select. If none of the combinations gives us the same information, then select the number 5 labeled none.

In the space provided tell why the items you marked in set I, II and III are correct. Use the words and symbols in the story.

12. Bob charged d cents for raking a lawn. He raked e lawns. He earned d x e cents for raking lawns.

CRITICAL THINKING

	I		II		III	
1.	d - e = d x e	_____	e + d = e + d	_____	$\frac{d \times e}{e} = d$	_____
2.	d x e = d x e	_____	e - d = e- d	_____	d x e = d	_____
3.	d x e = d + e	_____	d x d = e	_____	d x e = e x d	_____
4.	d / e = d x e	_____	d + e = d + e	_____	$\frac{d \times e}{d} = e$	_____
5.	none	_____	none	_____	none	_____

CRITICAL WRITING

____

____

____

____

____

____

____

CRITICAL READING

For each set (I, II, III) there is one or more correct answers to the story below. Which of the combinations gives us the same information found in the story? Place an x on the line after the response you select. If none of the combinations gives us the same information, then select the number 5 labeled none.

In the space provided tell why the items you marked in set I, II and III are correct. Use the words and symbols in the story.

13. Ann read one story d pages long. She read another story e pages long and a third story f pages long. Ann read d + e + f pages in all.

CRITICAL THINKING

	I		II		III	
1.	d x e x f = d + e + f	____	d + e + f = d + e + f	____	(d + e + f) - f = d + e	____
2.	d = d + e + f	____	(e + d) + f = e + d	____	d = e + f	____
3.	f + e + d = e + f + d	____	f + e = d	____	d x e = f	____
4.	f - d = d + e + f	____	d - e = f	____	d +e + f = e + f + d	____
5.	none	____	none	____	none	____

CRITICAL WRITING

____

____

____

____

____

____

____

____

CRITICAL READING

For each set (I, II, III) there is one or more correct answers to the story below. Which of the combinations gives us the same information found in the story? Place an x on the line after the response you select. If none of the combinations gives us the same information, then select the number 5 labeled none.

In the space provided tell why the items you marked in set I, II and III are correct. Use the words and symbols in the story.

14. Mother bought a melon for b cents and cookies for e cents and eggs for f cents. Mother spent b + e + f cents in all.

CRITICAL THINKING

	I		II		III	
1.	(b + e) + f = b + (e + f)	____	b + e + f = f + e + b	____	b x f = e	____
2.	b x e x f = b + e + f	____	b + e + f = b - e - f	____	b - e = f	____
3.	(b + e) - f = b + e + f	____	b +e + f = f x b	____	(b +e) - f = b + e + f	____
4.	b - e + f = b + e + f	____	b + f - e = b + f	____	b + e - f = b + f + e	____
5.	none	____	none	____	none	____

CRITICAL WRITING

____

____

____

____

____

____

____

____

CRITICAL READING

For each set (I, II, III) there is one or more correct answers to the story below. Which of the combinations gives us the same information found in the story? Place an x on the line after the response you select. If none of the combinations gives us the same information, then select the number 5 labeled none.

In the space provided tell why the items you marked in set I, II and III are correct. Use the words and symbols in the story.

15. In one section of the rodeo stand Jack counted d rows with e seats in each row. There were f sections in the rodeo stand. All sections were the same size. There were d x e x f seats in the rodeo stand.

CRITICAL THINKING

	I		II		III	
1.	d + e + f = d x e x f	____	f x e x d = d x f x e	____	d /e = e x d x f	____
2.	d x e x f = e x d x f	____	d + e +f = d +e +f	____	f /e = d	____
3.	d x e = f	____	d /e = f	____	e /f = d	____
4.	f + d - e = d x e x f	____	d / f = e	____	e x f x d = d x e x f	____
5.	none	____	none	____	none	____

CRITICAL WRITING

____

____

____

____

____

____

____

CRITICAL READING

For each set (I, II, III) there is one or more correct answers to the story below. Which of the combinations gives us the same information found in the story? Place an x on the line after the response you select. If none of the combinations gives us the same information, then select the number 5 labeled none.

In the space provided tell why the items you marked in set I, II and III are correct. Use the words and symbols in the story.

16. Howard bought a tablet for a cents, a ruler for b cents, and a drawing pencil for c cents. Howard spent a + b + c cents in all.

CRITICAL THINKING

	I		II		III	
1.	$a+b+c=b+c+a$	___	$(a+b+c)-c=a+b$	___	$c+b+a=a+b+c$	___
2.	$a+b+c=a \times b \times c$	___	$a \times b=c$	___	$a/b=c$	___
3.	$a-c=a+b+c$	___	$a+b+c=a+b+c$	___	$a=b \times c$	___
4.	$a=a+b+c$	___	$a/b+c=a+b+c$	___	$c-a/b=a \times b$	___
5.	none	___	none	___	none	___

CRITICAL WRITING

____

____

____

____

____

____

____

____

CRITICAL READING

For each set (I, II, III) there is one or more correct answers to the story below. Which of the combinations gives us the same information found in the story? Place an x on the line after the response you select. If none of the combinations gives us the same information, then select the number 5 labeled none.

In the space provided tell why the items you marked in set I, II and III are correct. Use the words and symbols in the story.

17. A DC-6 flew at an average speed of a miles an hour. It traveled b hours at this rate. The DC-6 flew a x b miles.

CRITICAL THINKING

	I		II		III	
1.	b x a = a x b	_____	a x b = a x b	_____	$\frac{b \times a}{b} = b$	_____
2.	b/a = a x b	_____	a /b = b x a	_____	b + a = a / b	_____
3.	b + a = a x b	_____	$\frac{a \times b}{a} = b$	_____	a + b = a = b	_____
4.	a - b = a x b	_____	a + b = b - a	_____	b x a = b x a	_____
5.	none	_____	none	_____	none	_____

CRITICAL WRITING

____

____

____

____

____

____

____

CRITICAL READING

For each set (I, II, III) there is one or more correct answers to the story below. Which of the combinations gives us the same information found in the story? Place an x on the line after the response you select. If none of the combinations gives us the same information, then select the number 5 labeled none.

In the space provided tell why the items you marked in set I, II and III are correct. Use the words and symbols in the story.

18. On a test, Sally had f examples right in addition. She had g examples right in subtraction, h right in multiplication, and i right in division. Sally had f + g + h + i right in all.

CRITICAL THINKING

	I		II		III	
1.	f+g+i=f+g+h+i	____	f+g=f+g	____	f-g=h+i	____
2.	fxgxhxi=f+g+h+i	____	f+g+h+i=i+f	____	f+h+g+i=g	____
3.	f+g+h+i=h+i	____	(f+g)=h-i	____	h+i=h+i	____
4.	f+g+i+h=h+i+f+g	____	f=g+h+I	____	h+f=g-i	____
5.	none	____	none	____	none	____

CRITICAL WRITING

____

____

____

____

____

____

____

CRITICAL READING

For each set (I, II, III) there is one or more correct answers to the story below. Which of the combinations gives us the same information found in the story? Place an x on the line after the response you select. If none of the combinations gives us the same information, then select the number 5 labeled none.

In the space provided tell why the items you marked in set I, II and III are correct. Use the words and symbols in the story.

19. Nan is c̲ years old. Her mother is f̲ years old. Nan is z̲ years younger than her mother.

CRITICAL THINKING

	I		II		III	
1.	c + f = z	_____	c + z = f	_____	c + f = z	_____
2.	f - c = z	_____	f /c = z	_____	z x c = f	_____
3.	c - f = z	_____	f x z = z	_____	f - z = c	_____
4.	z - c = f	_____	z / f = c	_____	c / f = z	_____
5.	none	_____	none	_____	none	_____

CRITICAL WRITING

____

____

____

____

____

____

____

____

CRITICAL READING

For each set (I, II, III) there is one or more correct answers to the story below. Which of the combinations gives us the same information found in the story? Place an x on the line after the response you select. If none of the combinations gives us the same information, then select the number 5 labeled none.

In the space provided tell why the items you marked in set I, II and III are correct. Use the words and symbols in the story.

20. Sam made e points in his first game of basketball. He made f points in the second game and g points in the third game. Sam made e + f + g points in all three games.

CRITICAL THINKING

	I		II		III	
1.	$e \times g = e + f + g$	_____	$(e + f + g) - (e + f) = g$	_____	$e \times g \times f = e \times g \times f$	_____
2.	$f + e + g = g + f + e$	_____	$e - f - g = e - f - g$	_____	$f / g = e + f + g$	_____
3.	$f - e - g = e + f + g$	_____	$g / f = e / g$	_____	$g - f = e + f + g$	_____
4.	$e \times f \times g = e + f + g$	_____	$e \times f = e / f$	_____	$(f + e + g) - e + g = f$	_____
5.	none	_____	none	_____	none	_____

CRITICAL WRITING

____

____

____

____

____

____

____

CRITICAL READING

For each set (I, II, III) there is one or more correct answers to the story below. Which of the combinations gives us the same information found in the story? Place an x on the line after the response you select. If none of the combinations gives us the same information, then select the number 5 labeled none.

In the space provided tell why the items you marked in set I, II and III are correct. Use the words and symbols in the story.

21. Carl cut <u>d</u> badges from a piece of ribbon. Each badge was <u>e</u> inches long. Carl used <u>d x e</u> inches of ribbon for the badges.

CRITICAL THINKING

	I		II		III	
1.	d x e = e x d	_____	d + e = d + e	_____	e / d = d x e	_____
2.	d - e = d x e	_____	e - d = e - d	_____	$\frac{d \times e}{e} = d$	_____
3.	d/e = d x e	_____	d x e = e x d	_____	d + e = d + e	_____
4.	e + d = e x d	_____	(d x e) e = d	_____	e - d = e	_____
5.	none	_____	none	_____	none	_____

CRITICAL WRITING

____

____

____

____

____

____

____

CRITICAL READING

For each set (I, II, III) there is one or more correct answers to the story below. Which of the combinations gives us the same information found in the story? Place an x on the line after the response you select. If none of the combinations gives us the same information, then select the number 5 labeled none.

In the space provided tell why the items you marked in set I, II and III are correct. Use the words and symbols in the story.

22. Mac picked b melons from his garden. He sold the melons for c cents apiece. He received b x c cents for the melons.

CRITICAL THINKING

	I		II		III	
1.	b + c = b x c	____	c x b = b x c	____	c - b = b /c	____
2.	b/c = b x c	____	b = c	____	b x c = b	____
3.	b x c = b - c	____	b - c = c/ b	____	$\frac{b \times c}{c} = b$	____
4.	b x c = b x c	____	b / c = b x c	____	b + c = b + c	____
5.	none	____	none	____	none	____

CRITICAL WRITING

____

____

____

____

____

____

____

CRITICAL READING

For each set (I, II, III) there is one or more correct answers to the story below. Which of the combinations gives us the same information found in the story? Place an x on the line after the response you select. If none of the combinations gives us the same information, then select the number 5 labeled none.

In the space provided tell why the items you marked in set I, II and III are correct. Use the words and symbols in the story.

23. a boys shared equally the c dollars they earned. Each boy received z dollars.

CRITICAL THINKING

	I		II		III	
1.	a + c = z	_____	a x z = c	_____	a x c = z	_____
2.	a - c = z	_____	a / c = z	_____	z - c = a	_____
3.	c/a = z	_____	z - a = c	_____	a + z = z	_____
4.	z/a = c	_____	a x a = z	_____	z x a = c	_____
5.	none	_____	none	_____	none	_____

CRITICAL WRITING

____

____

____

____

____

____

____

____

CRITICAL READING

For each set (I, II, III) there is one or more correct answers to the story below. Which of the combinations gives us the same information found in the story? Place an x on the line after the response you select. If none of the combinations gives us the same information, then select the number 5 labeled none.

In the space provided tell why the items you marked in set I, II and III are correct. Use the words and symbols in the story.

24. The flying distance from Chicago to Denver is e miles. The Jones family made the trip in f hours. The plane averaged z miles per hour.

CRITICAL THINKING

	I		II		III	
1.	f/e = z	_____	z x f = e	_____	f x e = z	_____
2.	f x e =z	_____	f /e = z	_____	e x f = z	_____
3.	e/f = z	_____	e + f = z	_____	f x z = e	_____
4.	e + f = z	_____	e - f = z	_____	z x e = f	_____
5.	none	_____	none	_____	none	_____

CRITICAL WRITING

____

____

____

____

____

____

____

____

CRITICAL READING

For each set (I, II, III) there is one or more correct answers to the story below. Which of the combinations gives us the same information found in the story? Place an x on the line after the response you select. If none of the combinations gives us the same information, then select the number 5 labeled none.

In the space provided tell why the items you marked in set I, II and III are correct. Use the words and symbols in the story.

25. Roger sold his pet white mice for a cents apiece. He sold b white mice. He received a x b for the mice.

CRITICAL THINKING

	I		II		III	
1.	a x b = a x b	_____	$\frac{a \times b}{b} = a$	_____	b /a = a x b	_____
2.	a + b = a x b	_____	$\frac{b \times a}{a} = b$	_____	b - a = a - b	_____
3.	b - a = a	_____	b - a = b - a	_____	a x b = b x a	_____
4.	a x b = b	_____	b x a = b	_____	b x a = b	_____
5.	none	_____	none	_____	none	_____

CRITICAL WRITING

____

____

____

____

____

____

____

CRITICAL READING

For each set (I, II, III) there is one or more correct answers to the story below. Which of the combinations gives us the same information found in the story? Place an x on the line after the response you select. If none of the combinations gives us the same information, then select the number 5 labeled none.

In the space provided tell why the items you marked in set I, II and III are correct. Use the words and symbols in the story.

26. Janet worked a arithmetic examples. b of them were addition. The rest were subtraction. Janet worked z subtraction examples.

CRITICAL THINKING

	I		II		III	
1.	a - b = z	_____	b + z = a	_____	b /z = a	_____
2.	a + b = z	_____	a x z = b	_____	a/z = b	_____
3.	a x b = z	_____	z x a = b	_____	a - z = b	_____
4.	a / b = z	_____	z / a = b	_____	a - b = a	_____
5.	none	_____	none	_____	none	_____

CRITICAL WRITING

____

____

____

____

____

____

____

____

CRITICAL READING

For each set (I, II, III) there is one or more correct answers to the story below. Which of the combinations gives us the same information found in the story? Place an x on the line after the response you select. If none of the combinations gives us the same information, then select the number 5 labeled none.

In the space provided tell why the items you marked in set I, II and III are correct. Use the words and symbols in the story.

27. Fred saved <u>b</u> dollars in <u>c</u> weeks. He saved <u>w</u> dollars each week.

CRITICAL THINKING

	I		II		III	
1.	w x b = c	_____	w x c = b	_____	w /c =d	_____
2.	b + c = w	_____	w / b = c	_____	c x w = b	_____
3.	b / c = w	_____	w - c = b	_____	c - w = b	_____
4.	c / w = b	_____	b + c = w	_____	c / b = w	_____
5.	none	_____	none	_____	none	_____

CRITICAL WRITING

____

____

____

____

____

____

____

____

CRITICAL READING

For each set (I, II, III) there is one or more correct answers to the story below. Which of the combinations gives us the same information found in the story? Place an x on the line after the response you select. If none of the combinations gives us the same information, then select the number 5 labeled none.

In the space provided tell why the items you marked in set I, II and III are correct. Use the words and symbols in the story.

28. Mrs. Nelson bought d chickens at e cents a pound. Each chicken weighed a pounds. Mrs. Nelson paid d x e x a cents for the chickens.

CRITICAL THINKING

	I		II		III	
1.	d x e = a	_____	d /e = a	_____	e / a = d	_____
2.	(d x e) x a = d x (e x a)	_____	a x d = e	_____	d + e = a	_____
3.	d + e + a = d x e x a	_____	d x e x a = e x d x a	_____	a / c = d	_____
4.	d - e = a	_____	d - e = a	_____	e - d = a	_____
5.	none	_____	none	_____	none	_____

CRITICAL WRITING

____

____

____

____

____

____

____

____

CRITICAL READING

For each set (I, II, III) there is one or more correct answers to the story below. Which of the combinations gives us the same information found in the story? Place an x on the line after the response you select. If none of the combinations gives us the same information, then select the number 5 labeled none.

In the space provided tell why the items you marked in set I, II and III are correct. Use the words and symbols in the story.

29. Mother baked f cherry tarts. The family ate e tarts for lunch. There were z tarts left.

CRITICAL THINKING

	I		II		III	
1.	z - f = e	_____	d + z + e = f + z + e	_____	z / f = e	_____
2.	f x z = e	_____	e + z = f	_____	f - z = e	_____
3.	z + e = f	_____	z x f = z x f	_____	z x e = f	_____
4.	e x f = z	_____	e - f = z	_____	f / e = z	_____
5.	none	_____	none	_____	none	_____

CRITICAL WRITING

____

____

____

____

____

____

____

____

CRITICAL READING

For each set (I, II, III) there is one or more correct answers to the story below. Which of the combinations gives us the same information found in the story? Place an x on the line after the response you select. If none of the combinations gives us the same information, then select the number 5 labeled none.

In the space provided tell why the items you marked in set I, II and III are correct. Use the words and symbols in the story.

30. Edward bought a box of b colored pencils for a cents. Each pencil cost z cents.

CRITICAL THINKING

	I		II		III	
1.	z/a = b	_____	a/z = b	_____	z + a = b	_____
2.	b x a = z	_____	b + a = z	_____	z x b = a	_____
3.	a/b = z	_____	a - b = z	_____	b x z = a	_____
4.	a - z = b	_____	z x b = z x b	_____	e x b = z	_____
5.	none	_____	none	_____	none	_____

CRITICAL WRITING

____

____

____

____

____

____

____

____

CRITICAL READING

For each set (I, II, III) there is one or more correct answers to the story below. Which of the combinations gives us the same information found in the story? Place an x on the line after the response you select. If none of the combinations gives us the same information, then select the number 5 labeled none.

In the space provided tell why the items you marked in set I, II and III are correct. Use the words and symbols in the story.

31. The Browns are taking a trip to the lake. The lake is a miles from home. They traveled b miles. They have w miles left to go.

CRITICAL THINKING

	I		II		III	
1.	a + b = w	_____	a / b = w	_____	b + w = a	_____
2.	a x b = w	_____	b - w = a	_____	b / a = w	_____
3.	w x b = a	_____	a - w = b	_____	w - a = b	_____
4.	b - w = a	_____	w + a = b	_____	a - b = w	_____
5.	none	_____	none	_____	none	_____

CRITICAL WRITING

____

____

____

____

____

____

____

____

CRITICAL READING

For each set (I, II, III) there is one or more correct answers to the story below. Which of the combinations gives us the same information found in the story? Place an x on the line after the response you select. If none of the combinations gives us the same information, then select the number 5 labeled none.

In the space provided tell why the items you marked in set I, and II and III are correct. Use the words and symbols in the story.

32. Lessons at the swimming pool cost a dollars a week. b children took lessons for c weeks. The lessons for the children cost a x b x c dollars.

CRITICAL THINKING

	I		II		III	
1.	(a x b) x c = a x (b x c)	_____	a x b x c = a x b x c	_____	b / a =c	_____
2.	a + b + c = a x b x c	_____	a x b = a x b	_____	a - c = b	_____
3.	a - c - b = a + b + c	_____	c/a = b	_____	a / b = c	_____
4.	a x b = c	_____	b - c = a	_____	b x c =a	_____
5.	none	_____	none	_____	none	_____

CRITICAL WRITING

____

____

____

____

____

____

____

CRITICAL READING

For each set (I, II, III) there is one or more correct answers to the story below. Which of the combinations gives us the same information found in the story? Place an x on the line after the response you select. If none of the combinations gives us the same information, then select the number 5 labeled none.

In the space provided tell why the items you marked in set I, II and III are correct. Use the words and symbols in the story.

33. The children took three spelling tests. Jack had a right on the first test. He had b right on the second test and c right on the third test. Jack had a + b + c right on all three tests.

CRITICAL THINKING

	I		II		III	
1.	a + (b + c) = (a + b) + c	____	c + b - a = c	____	c + a + b = b + c +a	____
2.	a + b + c = a x b x c	____	a (b + c) = a + b + c	____	c /a = b	____
3.	a - b - c = a + b + c	____	a x b x c = a x b x c	____	c (b + a) =a x b x c	____
4.	a + b + c = a + w + c	____	c + b + a = a + b	____	(c + b)- b = c	____
5.	none	____	none	____	none	____

CRITICAL WRITING

____

____

____

____

____

____

____

CRITICAL READING

For each set (I, II, III) there is one or more correct answers to the story below. Which of the combinations gives us the same information found in the story? Place an x on the line after the response you select. If none of the combinations gives us the same information, then select the number 5 labeled none.

In the space provided tell why the items you marked in set I, II and III are correct. Use the words and symbols in the story.

34. Janet worked c addition problems, d subtraction problems, and e division problems. Janet worked c + d + e problems in all.

CRITICAL THINKING

	I		II		III	
1.	c - d + e = c + d + e	____	(d + e) - e = d	____	c + d + e = c + d + e	____
2.	c + (d + e) = (c + d) + e	____	d / e = c	____	e + (d + c) = (e + d)	____
3.	c x d x e = c + d + e	____	c - d - e = c	____	c + d = d / e	____
4.	d + e = c + d + e	____	d x e = c / e	____	c + d+ e = d+ (e + c)	____
5.	none	____	none	____	none	____

CRITICAL WRITING

____

____

____

____

____

____

____

____

CRITICAL READING

For each set (I, II, III) there is one or more correct answers to the story below. Which of the combinations gives us the same information found in the story? Place an x on the line after the response you select. If none of the combinations gives us the same information, then select the number 5 labeled none.

In the space provided tell why the items you marked in set I, II and III are correct. Use the words and symbols in the story.

35. Alice weighs c pounds. Alice must gain z pounds before she will weigh d pounds.

CRITICAL THINKING

	I		II		III	
1.	c + d = z	_____	c + z = d	_____	c / z = d	_____
2.	d - c = z	_____	c + z = c + z	_____	z - c = d	_____
3.	c - z = d	_____	c x z = c x z	_____	z /c = d	_____
4.	c x z = d	_____	c + d = z - d	_____	d - z = c	_____
5.	none	_____	none	_____	none	_____

CRITICAL WRITING

____

____

____

____

____

____

____

____

CRITICAL READING

For each set (I, II, III) there is one or more correct answers to the story below. Which of the combinations gives us the same information found in the story? Place an x on the line after the response you select. If none of the combinations gives us the same information, then select the number 5 labeled none.

In the space provided tell why the items you marked in set I, II and III are correct. Use the words and symbols in the story.

36. Jim's class has raised a dollars from cookie sales. For b dollars, they bought a pair of hamsters, a pet pen for c dollars, and a tread wheel for d dollars. They spent b + c + d dollars in all.

CRITICAL THINKING

	I		II		III	
1.	$a + b + c + d = b + c + a$	____	$b + c + d = a$	____	$a - b = c + d$	____
2.	$a + b + c = b + c + d$	____	$b - c - d = a$	____	$b / d = a$	____
3.	$a = b + c + d$	____	$a / b = c + d$	____	$b / a = c$	____
4.	$(b + c) + d = b + (c + d)$	____	$d \times a = c$	____	$a - (b + c) = d$	____
5.	none	____	none	____	none	____

CRITICAL WRITING

____

____

____

____

____

____

____

CRITICAL READING

For each set (I, II, III) there is one or more correct answers to the story below. Which of the combinations gives us the same information found in the story? Place an x on the line after the response you select. If none of the combinations gives us the same information, then select the number 5 labeled none.

In the space provided tell why the items you marked in set I, II and III are correct. Use the words and symbols in the story.

37. Mike has d examples to do for his arithmetic lesson. He has worked c examples. Mike has z examples left to work.

CRITICAL THINKING

	I		II		III	
1.	c + d = z	_____	d - z = c	_____	d x z = c	_____
2.	c + z = d	_____	c /d = z	_____	d - z = c	_____
3.	z - c = d	_____	z + d + c	_____	d - c = z	_____
4.	z x d = e	_____	z / d = c	_____	z / e = d	_____
5.	none	_____	none	_____	none	_____

CRITICAL WRITING

____

____

____

____

____

____

____

____

CRITICAL READING

For each set (I, II, III) there is one or more correct answers to the story below. Which of the combinations gives us the same information found in the story? Place an x on the line after the response you select. If none of the combinations gives us the same information, then select the number 5 labeled none.

In the space provided tell why the items you marked in set I, II and III are correct. Use the words and symbols in the story.

38. James had <u>a</u> boxes. In each box he put <u>b</u> eggs. He sold the eggs for <u>c</u> cents apiece. He received <u>a x b x c</u> cents for the eggs.

CRITICAL THINKING

	I		II		III	
1.	a x b = c	_____	a - b = c	_____	a x c = b	_____
2.	a + b + c = a x b x c	_____	$\frac{a \times b}{a} = b$	_____	c x b = a	_____
3.	a x (b x c) = (a x b) x c	_____	a x b x c = a x b x c	_____	a x b = a x b	_____
4.	a - b - c = a x b	_____	a / b = c	_____	c x b x a = z	_____
5.	none	_____	none	_____	none	_____

CRITICAL WRITING

____

____

____

____

____

____

____

CRITICAL READING

For each set (I, II, III) there is one or more correct answers to the story below. Which of the combinations gives us the same information found in the story? Place an x on the line after the response you select. If none of the combinations gives us the same information, then select the number 5 labeled none.

In the space provided tell why the items you marked in set I, II and III are correct. Use the words and symbols in the story.

39. Carl cut a badges from a ribbon b inches long. The badges were all the same length. Each badge was z inches long.

CRITICAL THINKING

	I		II		III	
1.	a/b = z	_____	z x a = b	_____	z x b =a	_____
2.	a x b = z	_____	b/ z = a	_____	a / b = z	_____
3.	b/a = z	_____	a x z = b	_____	z + a = b	_____
4.	b - z = a	_____	a - b = z	_____	b - a = z	_____
5.	none	_____	none	_____	none	_____

CRITICAL WRITING

____

____

____

____

____

____

____

____

CRITICAL READING

For each set (I, II, III) there is one or more correct answers to the story below. Which of the combinations gives us the same information found in the story? Place an x on the line after the response you select. If none of the combinations gives us the same information, then select the number 5 labeled none.

In the space provided tell why the items you marked in set I, II and III are correct. Use the words and symbols in the story.

40. Farmer Nelson's hens laid f eggs one week. They laid a eggs the second week and b eggs the third week. The hens laid f + a + b eggs in the three weeks.

CRITICAL THINKING

	I		II		III	
1.	(f + a) + b = f + (a + b)	_____	f +a + b = f + a + b	_____	f +a = f + a	_____
2.	f - a - b = f + a + b	_____	f - a - b = f - a - b	_____	f / a = b	_____
3.	f + b = f x b	_____	f x b = b x f	_____	f + b = f + b	_____
4.	f + a + b = a x b	_____	f -a = b	_____	f + a + c = d	_____
5.	none	_____	none	_____	none	_____

CRITICAL WRITING

____

____

____

____

____

____

____

CRITICAL READING

For each set (I, II, III) there is one or more correct answers to the story below. Which of the combinations gives us the same information found in the story? Place an x on the line after the response you select. If none of the combinations gives us the same information, then select the number 5 labeled none.

In the space provided tell why the items you marked in set I, II and III are correct. Use the words and symbols in the story.

41. There are b desks in Bill's room. There are c desks in each row. There are z rows in the room.

CRITICAL THINKING

	I		II		III	
1.	b + c = z	_____	b /c = z	_____	z = c - b	_____
2.	z - c = b	_____	z x c = b	_____	z + b = c	_____
3.	c x z = b	_____	c / b = z	_____	b /z = c	_____
4.	c x b = z	_____	z / b = a	_____	z / b = c	_____
5.	none	_____	none	_____	none	_____

CRITICAL WRITING

____

____

____

____

____

____

____

____

CRITICAL READING

For each set (I, II, III) there is one or more correct answers to the story below. Which of the combinations gives us the same information found in the story? Place an x on the line after the response you select. If none of the combinations gives us the same information, then select the number 5 labeled none.

In the space provided tell why the items you marked in set I, II and III are correct. Use the words and symbols in the story.

42. Tickets for the puppet show were d cents apiece. e children sold f tickets each. They received d x e x f cents for the tickets sold.

CRITICAL THINKING

	I		II		III	
1.	d + e + f = d x e x f	____	d x e x f = d x e x f	____	d - f = e	____
2.	d - e = f	____	d / e =f	____	f - d = e	____
3.	d x (e x f) = (d x e) x f	____	d + e + f = f - d	____	e x f - d = a	____
4.	d x f x e = f	____	d x e = d x e	____	f / d = e	____
5.	none	____	none	____	none	____

CRITICAL WRITING

____

____

____

____

____

____

____

____

CRITICAL READING

For each set (I, II, III) there is one or more correct answers to the story below. Which of the combinations gives us the same information found in the story? Place an x on the line after the response you select. If none of the combinations gives us the same information, then select the number 5 labeled none.

In the space provided tell why the items you marked in set I, II and III are correct. Use the words and symbols in the story.

43. Joe's father paid e cents for f tickets to the basketball game. He spent e x f cents for the tickets.

CRITICAL THINKING

	I		II		III	
1.	e /f = e x f	_____	e x f = f - e	_____	e +f = e + f	_____
2.	e + f = e x f	_____	$\frac{f \times e}{e} = f$	_____	$\frac{e \times f}{f} = e$	_____
3.	e x f = f x e	_____	f + e = f / e	_____	f - e = f - e	_____
4.	e x f = f	_____	f / e = w	_____	e - f = e / f	_____
5.	none	_____	none	_____	none	_____

CRITICAL WRITING

____

____

____

____

____

____

____

CRITICAL READING

For each set (I, II, III) there is one or more correct answers to the story below. Which of the combinations gives us the same information found in the story? Place an x on the line after the response you select. If none of the combinations gives us the same information, then select the number 5 labeled none.

In the space provided tell why the items you marked in set I, II, and III are correct. Use the words and symbols in the story.

44. Miss Smith gave each of e children a sheet of paper. On each sheet were b rows of c squares each. There were e x b x c squares in the sheets of paper.

CRITICAL THINKING

	I		II		III	
1.	(e x b) x c = e x (b x c)	_____	e / b + c = e x b	_____	b x c = b x c	_____
2.	e x c = e x b x c	_____	e x b x c = c / b	_____	e /b = b / e	_____
3.	e + b + c = e x b x c	_____	e + b = c	_____	e - c = b	_____
4.	$\frac{e - b}{c}$ = e x b x c	_____	e x b x c = e x b x d	_____	b / c + e = c	_____
5.	none	_____	none	_____	none	_____

CRITICAL WRITING

____

____

____

____

____

____

CRITICAL READING

For each set (I, II, III) there is one or more correct answers to the story below. Which of the combinations give us the same information found in the story? Place an x on the line after the response you select. If none of the combinations gives us the same information then select the number 5 labeled none.

In the space provided tell why the item's you marked in set I II and III, are correct. Use the words and symbols in the story

45. A lady buys d peppers for e cents. All the peppers are the same price. Each cost z cents.

CRITICAL THINKING

	I		II		III	
1.	z x e = d	_____	e / z = d	_____	d + e = e + d	_____
2.	e x d = z	_____	e - z = d / e	_____	z + e = d - e	_____
3.	d x z = e	_____	d + e + z = z - d	_____	d / e = z	_____
4.	d + e = z	_____	e - z = d	_____	d - z = e + d	_____
5.	none	_____	none	_____	none	_____

CRITICAL WRITING

____

____

____

____

____

____

____

____

CRITICAL READING

For each set (I, II, III) there is one or more correct answers to the story below. Which of the combinations gives us the same information found in the story? Place an x on the line after the response you select. If none of the combinations gives us the same information, then select the number 5 labeled none.

In the space provided tell why the items you marked in set I, II and III are correct. Use the words and symbols in the story.

46. Bill bought a tops at b cents apiece. All the tops cost the same. Bill paid a x b cents for the tops.

CRITICAL THINKING

	I		II		III	
1.	a + b = a x b	_____	a / b = b	_____	a x b = b x a	_____
2.	a x b = a x b	_____	a + b = a + b	_____	a - b = a	_____
3.	b - a = a x b	_____	b x a = b / a	_____	a + b = b x a	_____
4.	a / b = a x b	_____	$\frac{a \times b}{b} = a$	_____	b x a = b	_____
5.	none	_____	none	_____	none	_____

CRITICAL WRITING

____

____

____

____

____

____

____

CRITICAL READING

For each set (I, II, III) there is one or more correct answers to the story below. Which of the combinations gives us the same information found in the story? Place an x on the line after the response you select. If none of the combinations gives us the same information, then select the number 5 labeled none.

In the space provided tell why the items you marked in set I, II and III are correct. Use the words and symbols in the story.

47. Kenneth sells tomatoes for d cents a box. He packs f tomatoes in each box. Each tomato costs w cents.

CRITICAL THINKING

	I		II		III	
1.	d x w = f	_____	d x f = w	_____	d + e = w	_____
2.	f - w = d	_____	w - f = d	_____	d - w = d - w	_____
3.	d x f = w	_____	f + w = d	_____	f x w = d	_____
4.	w x f = d	_____	f x w = d	_____	d / w = f	_____
5.	none	_____	none	_____	none	_____

CRITICAL WRITING

____

____

____

____

____

____

____

____

CRITICAL READING

For each set (I, II, III) there is one or more correct answers to the story below. Which of the combinations gives us the same information found in the story? Place an x on the line after the response you select. If none of the combinations gives us the same information, then select the number 5 labeled none.

In the space provided tell why the items you marked in set I, II and III are correct. Use the words and symbols in the story.

48. Nick has a pictures of baseball players. He trades b of them with Andy for a knife. Nick had z baseball pictures left.

CRITICAL THINKING

	I		II		III	
1.	z + b = a	_____	a - z = b	_____	a / b = z	_____
2.	z + a = b	_____	a / b = a / b	_____	a + b =a + b	_____
3.	z - a = b	_____	b + a = a + b	_____	a - b - z = a + b	_____
4.	a x b = z	_____	z x a = b	_____	a - b = z	_____
5.	none	_____	none	_____	none	_____

CRITICAL WRITING

____

____

____

____

____

____

____

____

CRITICAL READING

For each set (I, II, III) there is one or more correct answers to the story below. Which of the combinations gives us the same information found in the story? Place an x on the line after the response you select. If none of the combinations gives us the same information, then select the number 5 labeled none.

In the space provided tell why the items you marked in set I, II and III are correct. Use the words and symbols in the story.

49. a children wished to share equally b candy sticks. Each child received w candy sticks.

CRITICAL THINKING

	I		II		III	
1.	b - a = w	_____	w x a = b	_____	w x b = a	_____
2.	w = b x a	_____	b / a = w	_____	b/w = a	_____
3.	w + b = a	_____	a + b = w	_____	a / w = b	_____
4.	a x w = b	_____	w - a = b	_____	b - w = a	_____
5.	none	_____	none	_____	none	_____

CRITICAL WRITING

____

____

____

____

____

____

____

____

CRITICAL READING

For each set (I, II, III) there is one or more correct answers to the story below. Which of the combinations gives us the same information found in the story? Place an x on the line after the response you select. If none of the combinations gives us the same information, then select the number 5 labeled none.

In the space provided tell why the items you marked in set I, II and III are correct. Use the words and symbols in the story.

50. The children bought a plants for window boxes for the school. It took b plants to fill each box. The children filled z window boxes.

CRITICAL THINKING

	I		II		III	
1.	b - a = z	_____	z / b = a	_____	z + a =b	_____
2.	z x b = a	_____	b / a = z	_____	a - z = b	_____
3.	a x z = b	_____	a + b = a + b	_____	a x z = b	_____
4.	a + b = z	_____	z - b = a	_____	z + a = b x a	_____
5.	none	_____	none	_____	none	_____

CRITICAL WRITING

____

____

____

____

____

____

____

____

CRITICAL READING

For each set (I, II, III) there is one or more correct answers to the story below. Which of the combinations gives us the same information found in the story? Place an x on the line after the response you select. If none of the combinations gives us the same information, then select the number 5 labeled none.

In the space provided tell why the items you marked in set I, II and III are correct. Use the words and symbols in the story

51. *THE TRIP TO GRAND CITY*

Two weeks before Christmas, the Ross Family went to Grand City to sell Christmas trees and homemade jelly. Mr. Ross and Ray went in the truck and Mrs. Ross and Ruth used the family car. The trip to Grand City was $\underline{x}$ miles.

On the way to and from Grand City, Mr. Ross used $\underline{a}$ gallons of gasoline and $\underline{b}$ quarts of oil. The gasoline cost $\underline{c}$ dollars per gallon and oil cost $\underline{d}$ dollars per quart. Mrs. Ross used $\underline{e}$ gallons of gasoline at $\underline{c}$ dollars per gallon. The gasoline they used cost $\underline{n}$ dollars, and the oil they used cost $\underline{w}$ dollars. They spent $\underline{y}$ dollars on gasoline and oil for the trip. The cost per mile was $\underline{z}$ dollars.

CRITICAL THINKING

	I		II		III	
1.	$z = (a \times c) + (e \times c) + b$	_____	$y = n + w$	_____	$y - n = w$	_____
2.	$z = \frac{x - c + a}{e}$	_____	$x = \frac{a + b}{z}$	_____	$\frac{w + y}{z} = x$	_____
3.	$z = \frac{(d + c) + (e \times d)}{x}$	_____	$y = \frac{x + z}{n}$	_____	$c \times e = e \times c$	_____
4.	$z = \frac{(a \times c) + (e \times c) + (b \times d)}{x}$	_____	$\frac{(a \times c) + b}{w} = z$	_____	$\frac{y}{x} = z$	_____
5.	none	_____	none	_____	none	_____

CRITICAL WRITING

CRITICAL READING

For each set (I, II, III) there is one or more correct answers to the story below. Which of the combinations gives us the same information found in the story? Place an x on the line after the response you select. If none of the combinations gives us the same information, then select the number 5 labeled none.

In the space provided tell why the items you marked in set I, II and III are correct. Use the words and symbols in the story.

52. Billy started a bicycle ride to the store at home (Xo). He started at time (To) and arrived at the store at a distance (Xf) at time (Tf). His average speed was (V), therefore:

CRITICAL THINKING

	I		II		III	
1.	Xf - Tf = V	____	V x (Tf - To) = Xf - Xo	____	$\frac{Xf - Xo}{V}$ = Tf - To	____
2.	$\frac{Tf - To}{Xf - To}$ = V	____	Tf - To = Xf - X o	____	$\frac{X}{T}$ = V	____
3.	$\frac{Xf - Xo}{Tf - To}$ = V	____	V x T = X	____	$\frac{X}{V}$ = T	____
4.	$\frac{Xf - Tf}{To - Xo}$ = V	____	Tf - Xo = Xf	____	To - V = Xo	____
5.	none	____	none	____	none	____

CRITICAL WRITING

____

____

____

____

____

CRITICAL READING

For each set (I, II, III) there is one or more correct answers to the story below. Which of the combinations gives us the same information found in the story? Place an x on the line after the response you select. If none of the combinations gives us the same information, then select the number 5 labeled none.

In the space provided tell why the items you marked in set I, II and III are correct. Use the words and symbols in the story.

53. When you ride your bike at the same speed (V) for hours (T), you travel (X) miles. Thus:

CRITICAL THINKING

	I		II		III	
1.	$\frac{V}{T} = X$	_____	$T \times X = V$	_____	$\frac{X}{T} = V$	_____
2.	$X - V = T$	_____	$T = \frac{V}{X}$	_____	$X + T = V$	_____
3.	$T \times V = X$	_____	$X - T = V$	_____	$X + T = V$	_____
4.	$V \times T = X$	_____	$T \times T = X \times X$	_____	$V - T = X$	_____
5.	none	_____	none	_____	none	_____

CRITICAL WRITING

____

____

____

____

____

____

____

CRITICAL READING

For each set (I, II, III) there is one or more correct answers to the story below. Which of the combinations gives us the same information found in the story? Place an x on the line after the response you select. If none of the combinations gives us the same information, then select the number 5 labeled none.

In the space provided tell why the items you marked in set I, II and III are correct. Use the words and symbols in the story.

54. When traveling, you had to speed up to (Vo) at time (To) and return to the same speed (Vf) at time (Tf). Your change in speed (A) was :

CRITICAL THINKING

	I		II		III	
1.	Vf + To = A	_____	A x (To -Tf) = Vo-Vf	_____	$\frac{To - Tf}{A}$ = Vo-Vf	_____
2.	Vf x Vo x So = A	_____	A - To = Tf - Vo	_____	$\frac{To}{A}$ = Vf	_____
3.	$\frac{Tf - To}{Vf - Vo}$ = A	_____	Vf + Vo = Tf +	_____	A + To = Vo	_____
4.	$\frac{Vo - Vf}{To - Tf}$ = A	_____	$\frac{Vf}{A}$ = To	_____	A x Vf x To = o	_____
5.	none	_____	none	_____	none	_____

CRITICAL WRITING

____

____

____

____

____

____

CRITICAL READING

For each set (I, II, III) there is one or more correct answers to the story below. Which of the combinations gives us the same information found in the story? Place an x on the line after the response you select. If none of the combinations gives us the same information, then select the number 5 labeled none.

In the space provided tell why the items you marked in set I, II and III are correct. Use the words and symbols in the story.

55. To cross a river in your boat in (T) minutes at a speed of (Vbg)miles per hour when the water is moving south at a speed of (Vwg) miles per hour, your boat will land (Dwg) miles down river.

CRITICAL THINKING

	I		II		III	
1.	Dwg = Vwg + Vwg	_____	$Vwg = \frac{Dwg}{T}$	_____	T = Dwg x Vwg	_____
2.	Dwg = Vbg x T	_____	Vwg = Dwg x T	_____	T = Vwg x Dwg	_____
3.	Dwq = Vbg - Vwg	_____	Dwg = Dwg x T	_____	$Dwg = \frac{Vwg}{T}$	_____
4.	D wg= Vwg x T	_____	T = Dwg + Vwg	_____	$T = \frac{Dwg}{Vwg}$	_____
5.	none	_____	none	_____	none	_____

CRITICAL WRITING

____

____

____

____

____

____

About the Author

PEOPLE. Robert Hammond	WALL OF HONOR The Ohio State Years

Robert Hammond came to Ohio State from the University of Arizona, where he headed the Epic Evaluation Center, to be associate director of The Evaluation Center. Dr. Hammond was an incredibly influential and forceful advocate for and leader of teachers and other persons working in schools. He organized a consortium of small schools in Arizona that became a collaborative force for effective evaluation of federal projects. He was instrumental in linking the Center to the realities of schools and school districts, especially small rural districts.

978-0-595-40784-2
0-595-40784-6

www.ingramcontent.com/pod-product-compliance
Lightning Source LLC
Chambersburg PA
CBHW030800180526
45163CB00003B/1108

* 9 7 8 0 5 9 5 4 0 7 8 4 2 *